Name ______________________________

Moon, Earth, and Sun

A star is a ball of burning gases.
The sun is a star.
A planet goes around the sun.
Earth is a planet.
A moon is a satellite (sat-e-lite) that goes around a planet.
Earth has one moon.

Match:

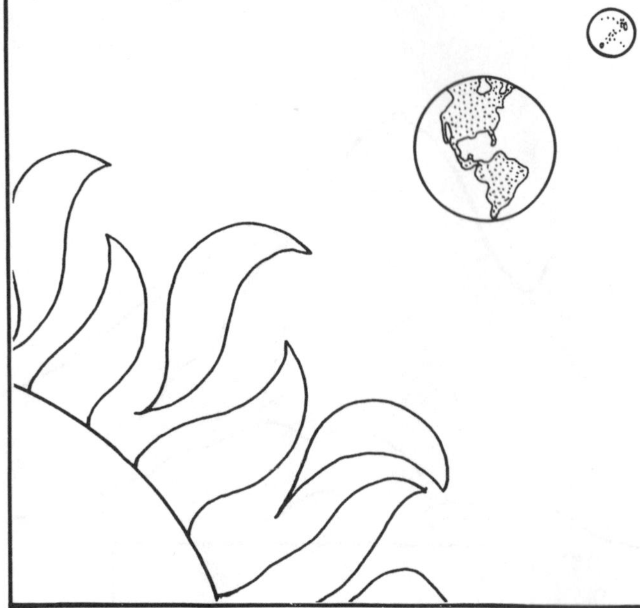

- earth
- moon
- sun

Fill in the blanks:

1. The moon is a ____________.
2. The earth is a ____________.
3. The sun is a ____________.

satellite
star
planet

Extra: Put an X on the picture that shows where you live.

Teacher—You may want to introduce the terms orbit, satellite and reflection.

Name ______________________________

The Moon

The moon is smaller than the earth and sun. We can see the moon because it shines with light from the sun. The moon goes around the earth one time in a month. There is no air or water on the moon. There are tall mountains and flat, dusty plains on the moon. There are many big holes called craters (kray-terz) on the moon.

plain	mountain	crater

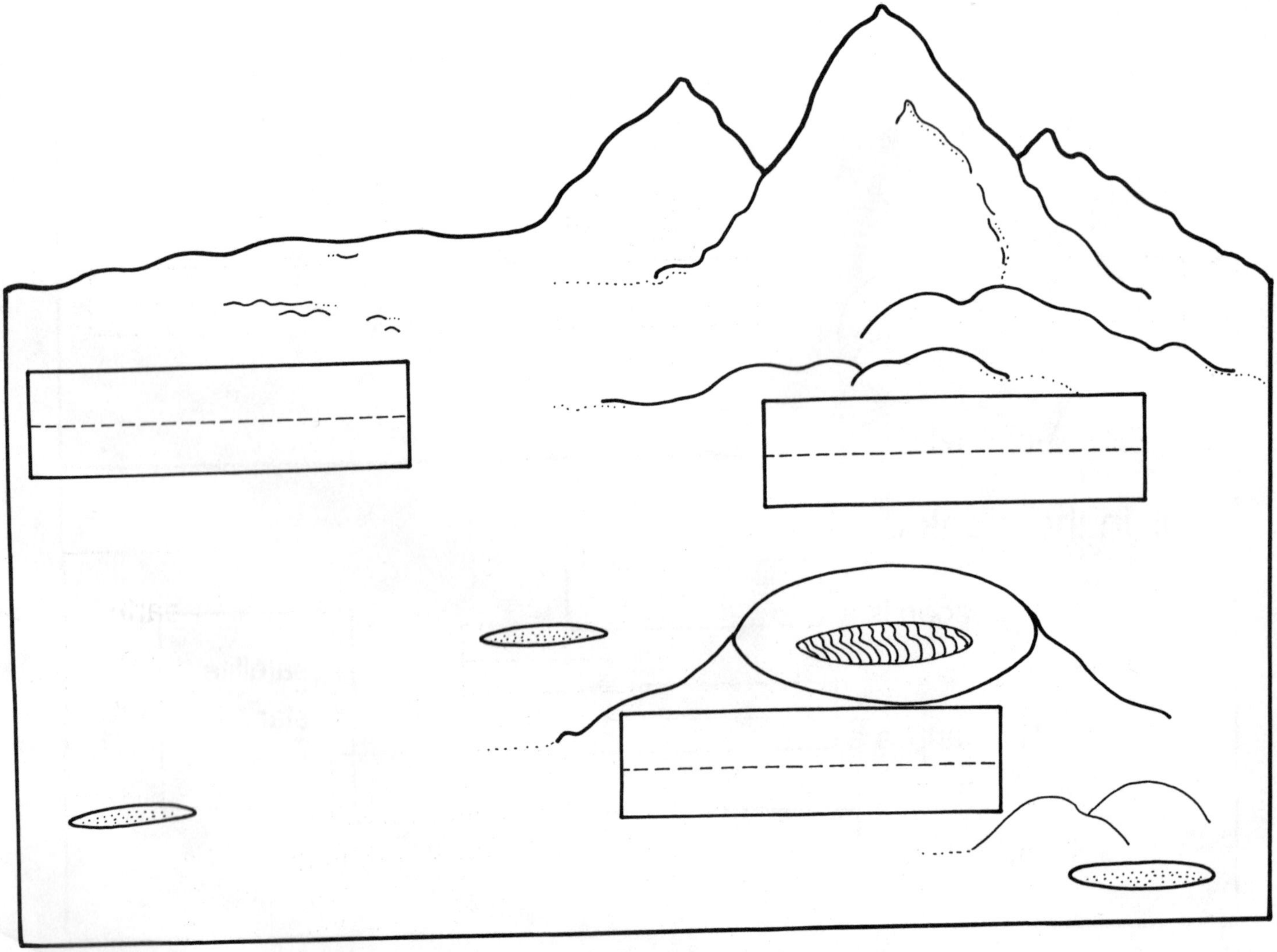

Extra: Make the earth in the sky above the moon.

Name ____________________

See The Moon Change

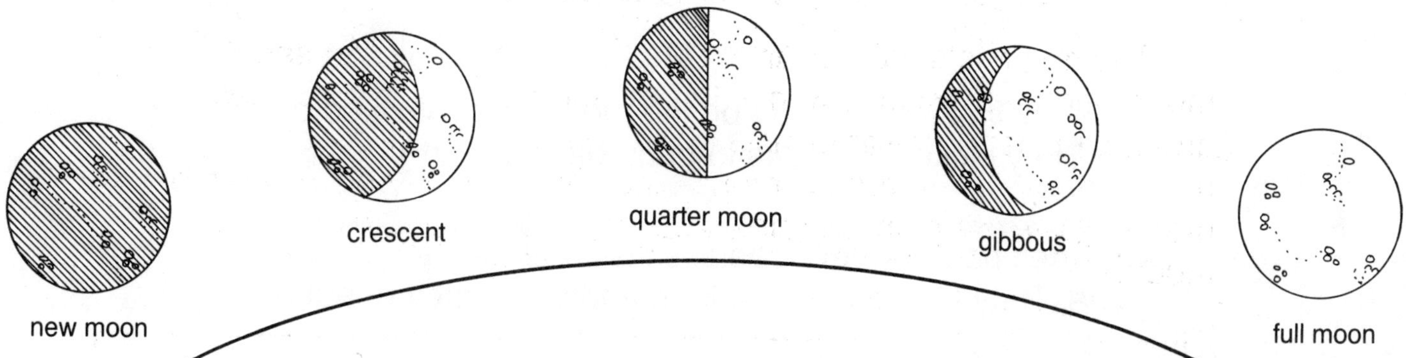

We see the moon at night.
The moon gets its light from the sun.
The moon is always round, but we do not see all of it every night.
As the moon goes around the earth, it looks different to us.

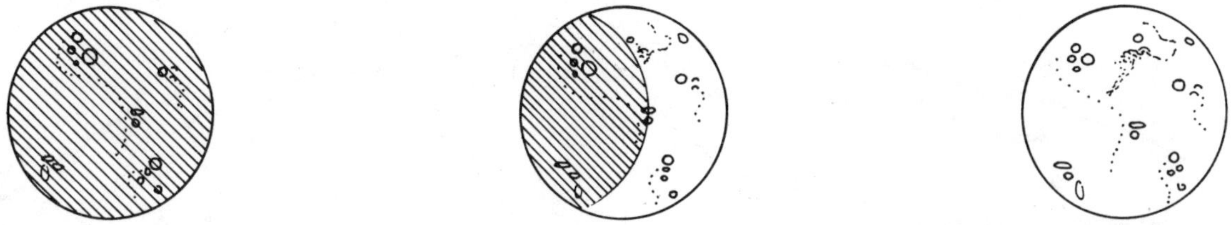

[______]   [______]   [______]

Fill in the blanks:

1. We see the moon at [______] .
2. The moon gets light from the [______] .
3. The [______] goes around the [______] .

earth
moon
night
sun

Extra: Draw a quarter moon on the back of this paper.

Teacher: You may want to discuss the journey of Neil Armstrong and Buzz Aldrin.

Name ____

Men On the Moon

In 1969 the first men landed on the moon. They walked on the moon. They picked up rocks to bring back to earth.

These men had to wear special suits. They had to take air to breathe. They had to take the food and water they needed for the trip.

Make a man on this moon.
What will he have on?

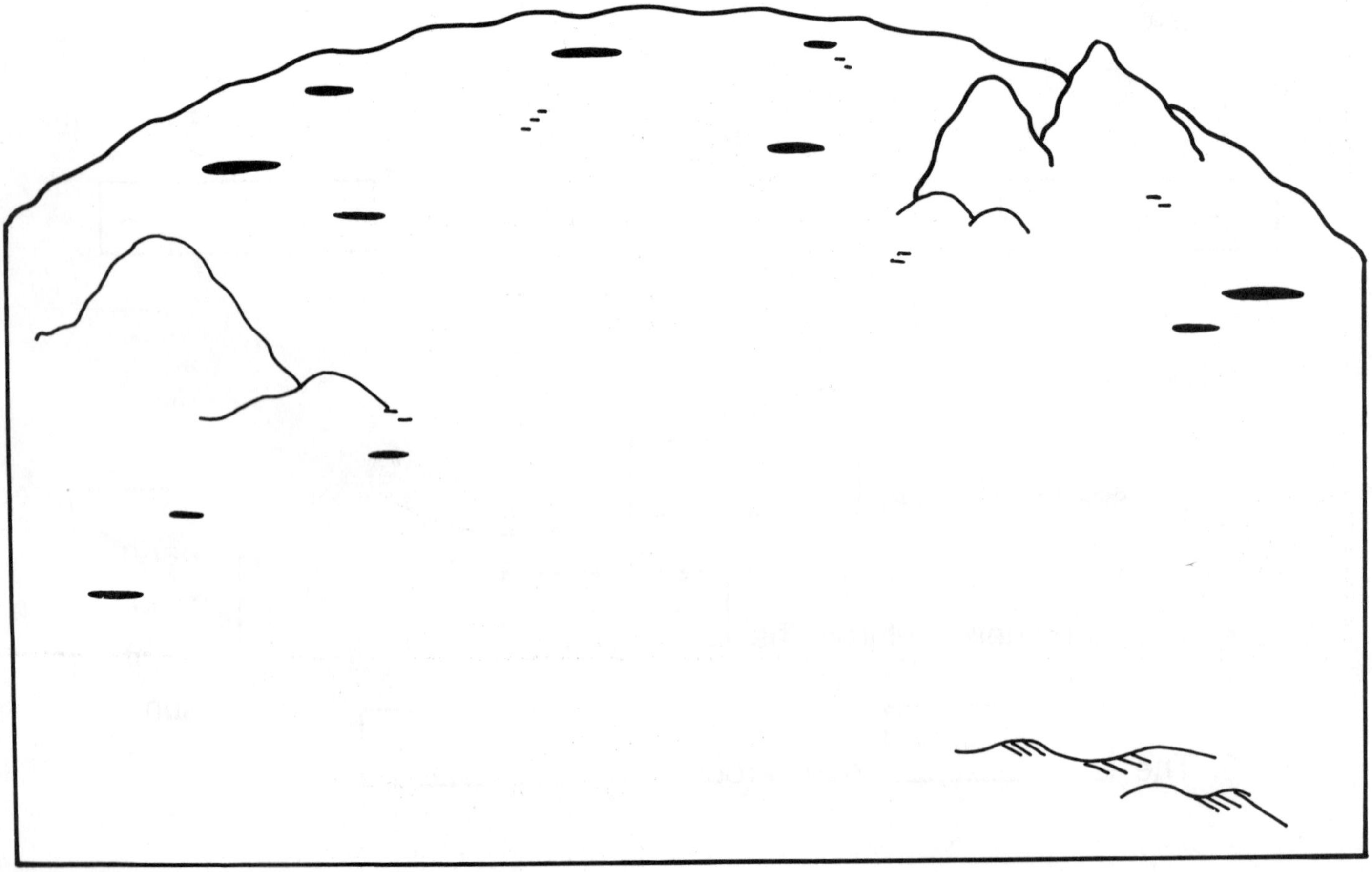

Extra: How are the earth and the moon alike?
How are they different?

Teacher: You may want to introduce these terms: rotate, orbit.

Name ______________________________________

Sun

The sun is a star. It is the star closest to the earth. It is the star we see in the daytime. The sun is much bigger than the earth. It is so far away, it looks small. The earth goes around the sun. It takes one year.

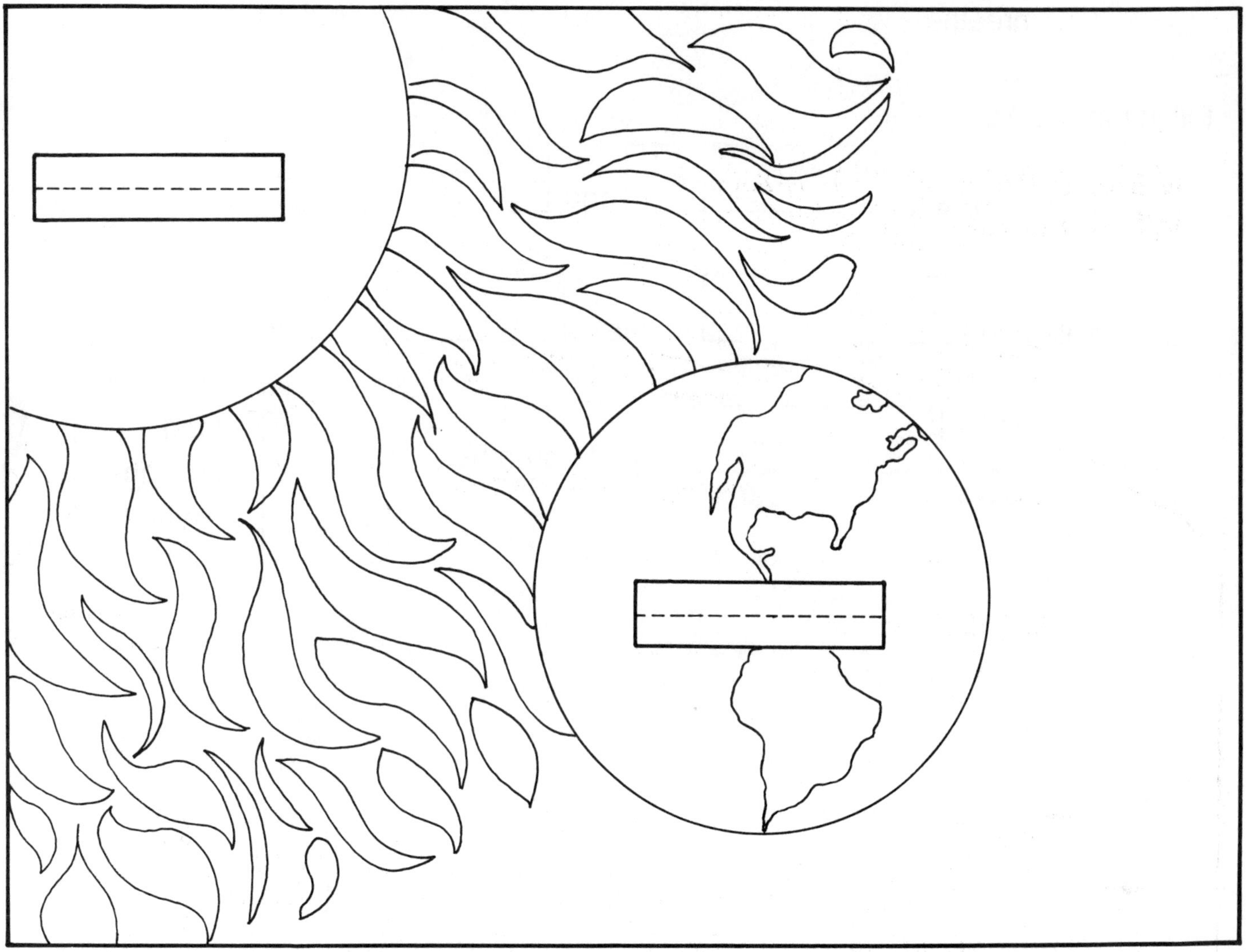

Color the sun yellow.
Color the earth blue and green.

Extra: The moon goes around the earth. Make a little moon by the earth.

Teacher: You may want to discuss solar energy at this time.

Name ____________________________________

We Need the Sun

The sun is a ball of hot gases. It gives off heat and light.

We need the sunlight. It gives us heat to stay warm. It gives us light to see by. Plants need the sunlight too. It helps plants make food for us to eat and oxygen (ok-si-jen) for us to breathe.

Fill in the blanks:

1. The sun gives us __________ and __________.

2. Plants need __________ to make __________ for us to eat.

3. Plants make __________ for us to breathe.

food
sunlight
oxygen
light
heat

Draw your backyard on a sunny day.

Circle the things that need sunshine.

Extra: Get a sheet of paper. Tell one way the sun helps us.

Name________________________________

Night and Day

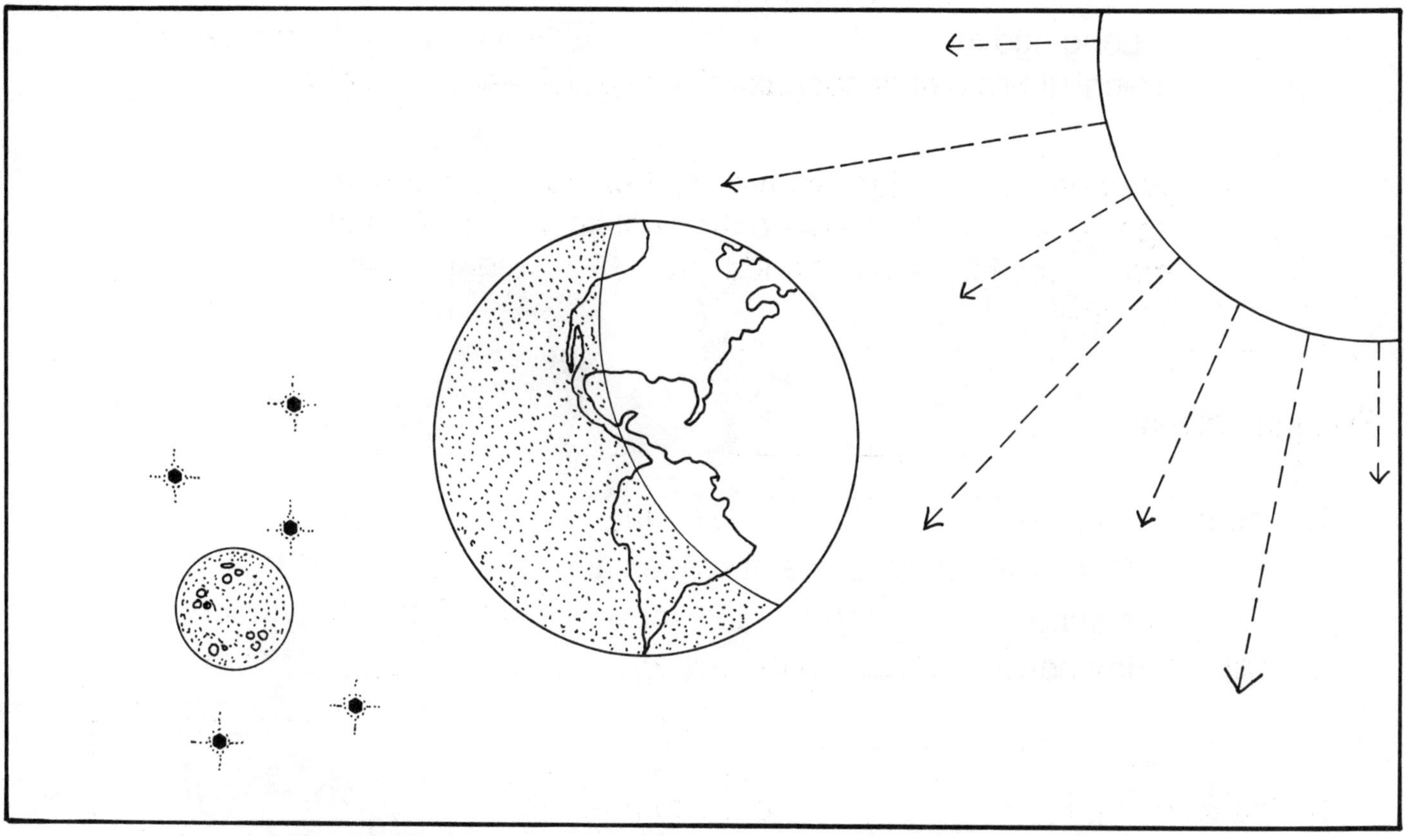

The sun shines all the time.
The earth turns so we cannot see the sun's light at night.
Light fills the sky in the daytime.
The sky is dark at night.

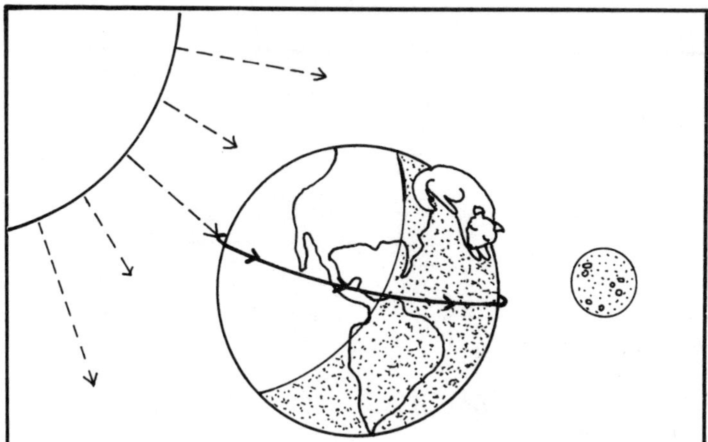

Extra: Color the night side black.
Color the day side yellow.

Name____

A Sunny Way to Tell Time

Long ago the sun helped show what time it was. When sunlight hits something, it makes a shadow.

Shadows move as the sun goes across the sky. Men made a <u>*sundial*</u>. When the shadow moved, it would touch numbers to show the time.

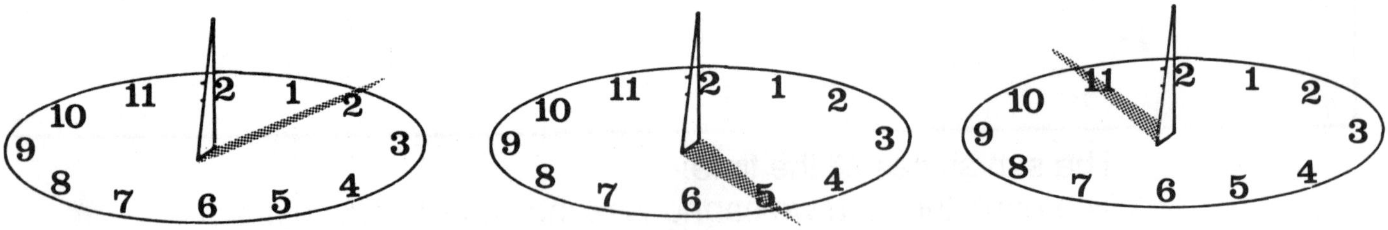

A sundial was no help on a rainy day or at night!

What time is it?

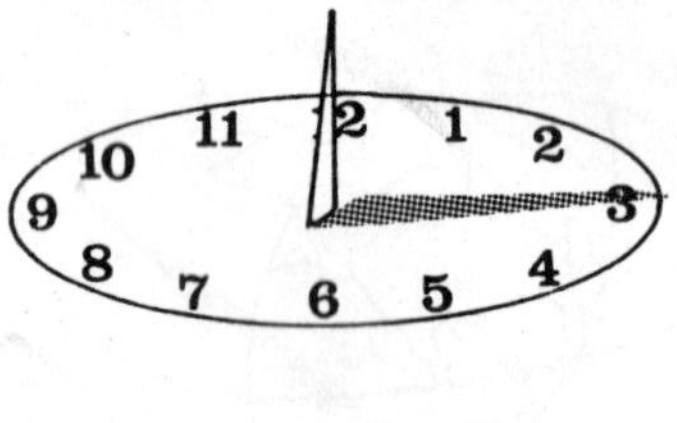 ☐ o'clock	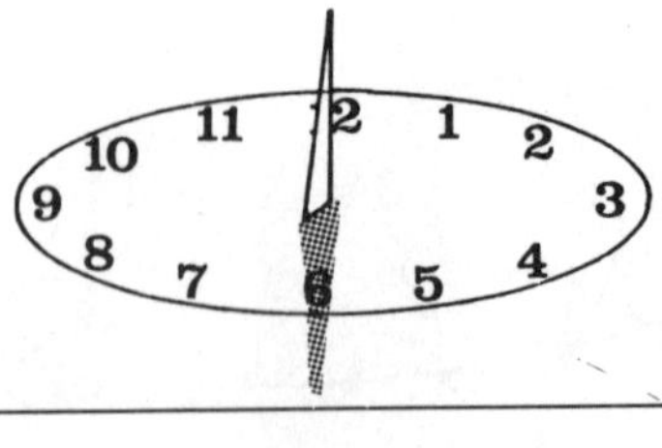 ☐ o'clock
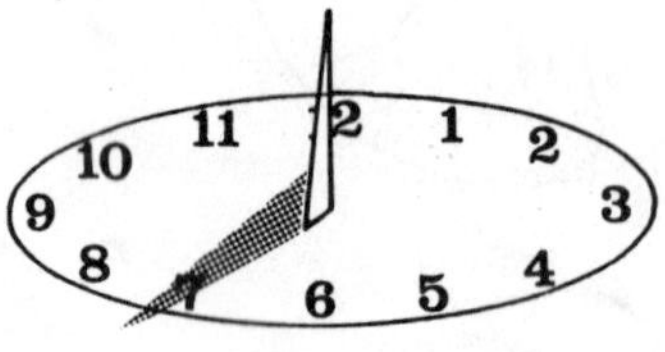 ☐ o'clock	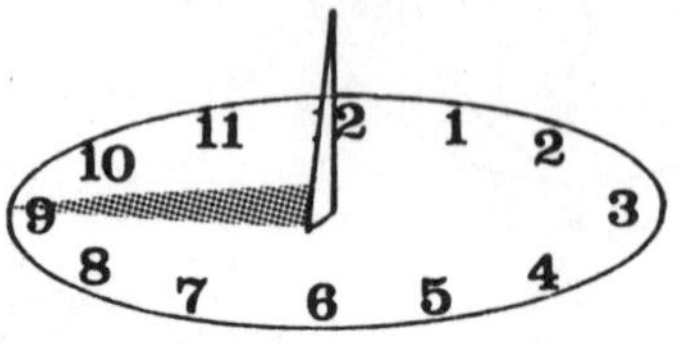 ☐ o'clock

Extra: Show what the hour is in your classroom.

Name ____________________

Eclipse (ee-klips)

What happens when the earth's shadow falls on the moon? The moon cannot get light from the sun, so you cannot see the moon.

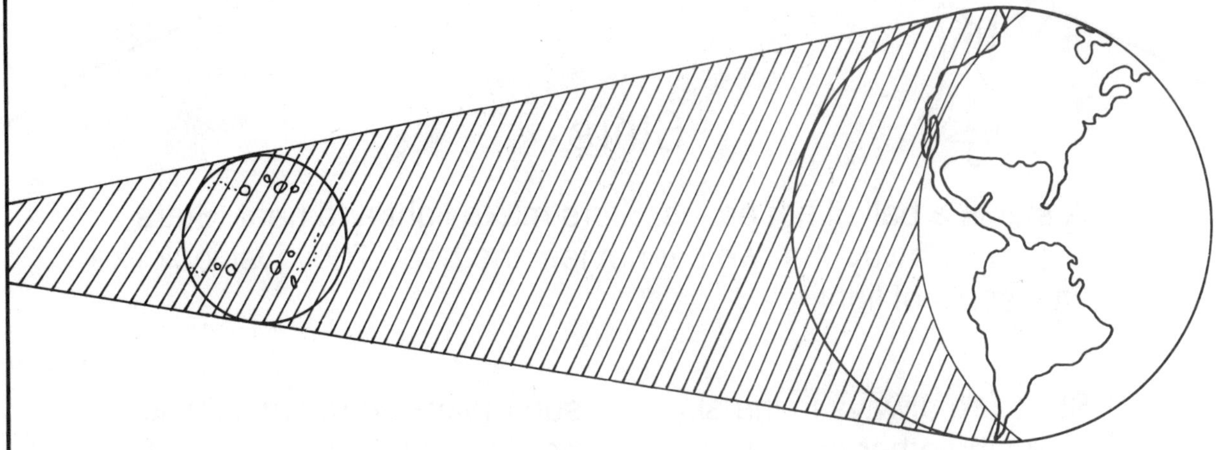

eclipse of the moon

What happens when the moon's shadow falls on the earth? The moon blocks the sun's light from part of the earth, so that part cannot see the sun.

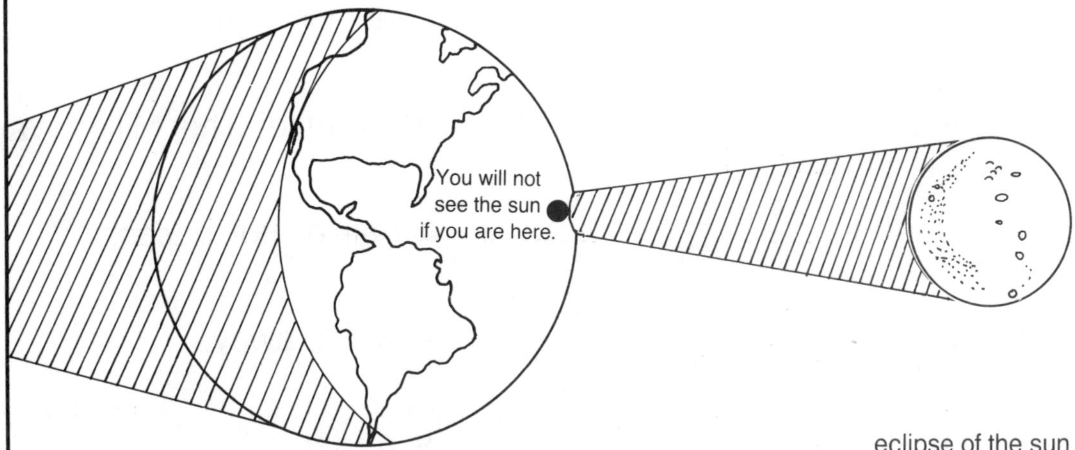

eclipse of the sun

Extra: Ask 10 people "Have you ever seen an eclipse?"

Teacher: You may want to introduce the concept of light years.

Name ______________________________

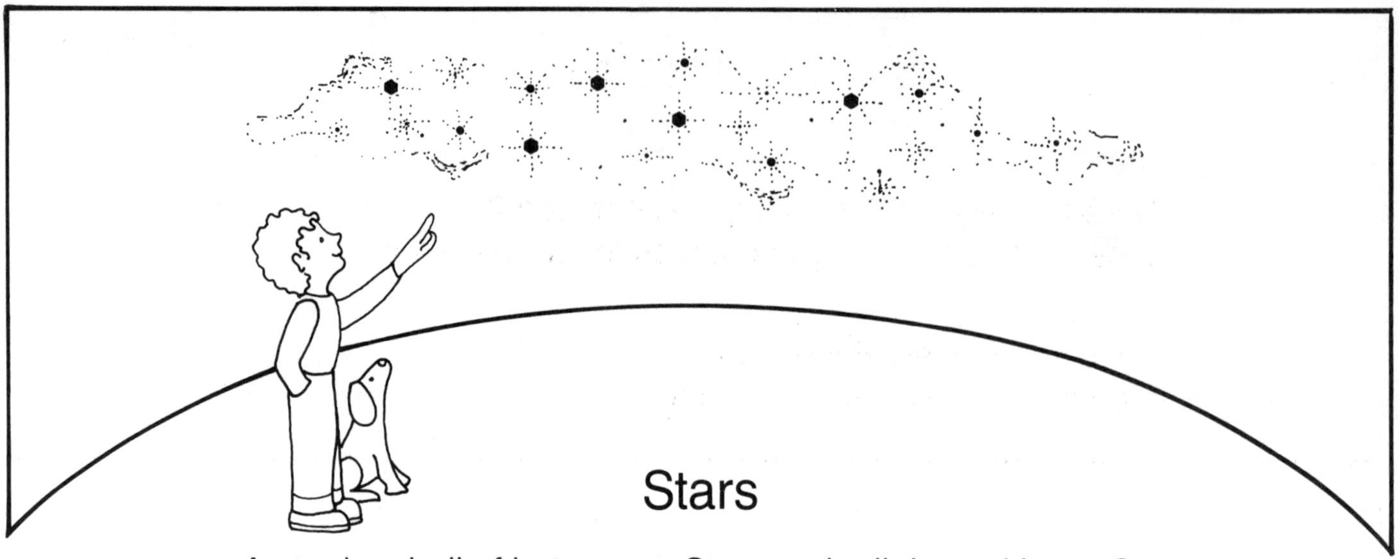

Stars

A star is a ball of hot gases. Stars make light and heat. Stars are so far away that we cannot feel the heat. They just look like little spots of light.

Stars are always in the sky. Our sun makes so much light, it hides the other stars in the daytime. At night, when the sun is not in our sky, we can see the other stars.

The stars in our sky are called "The Milky Way". Can you guess why?

Match:

1. A star is	heat and light.
2. Stars are	are part of The Milky Way.
3. A star makes	a ball of hot gases.
4. Our sun	very far away.
5. The stars in our sky	is a star.

Extra: What is the name of the star nearest the earth?

Teacher: You may want to teach the term constellation.

Name ______________________________

Sky Pictures

The night sky is full of stars. You can see 2000 stars with your eyes. Long ago men gave names to groups of stars.

Have you looked at the night sky?
Did you see these star pictures?

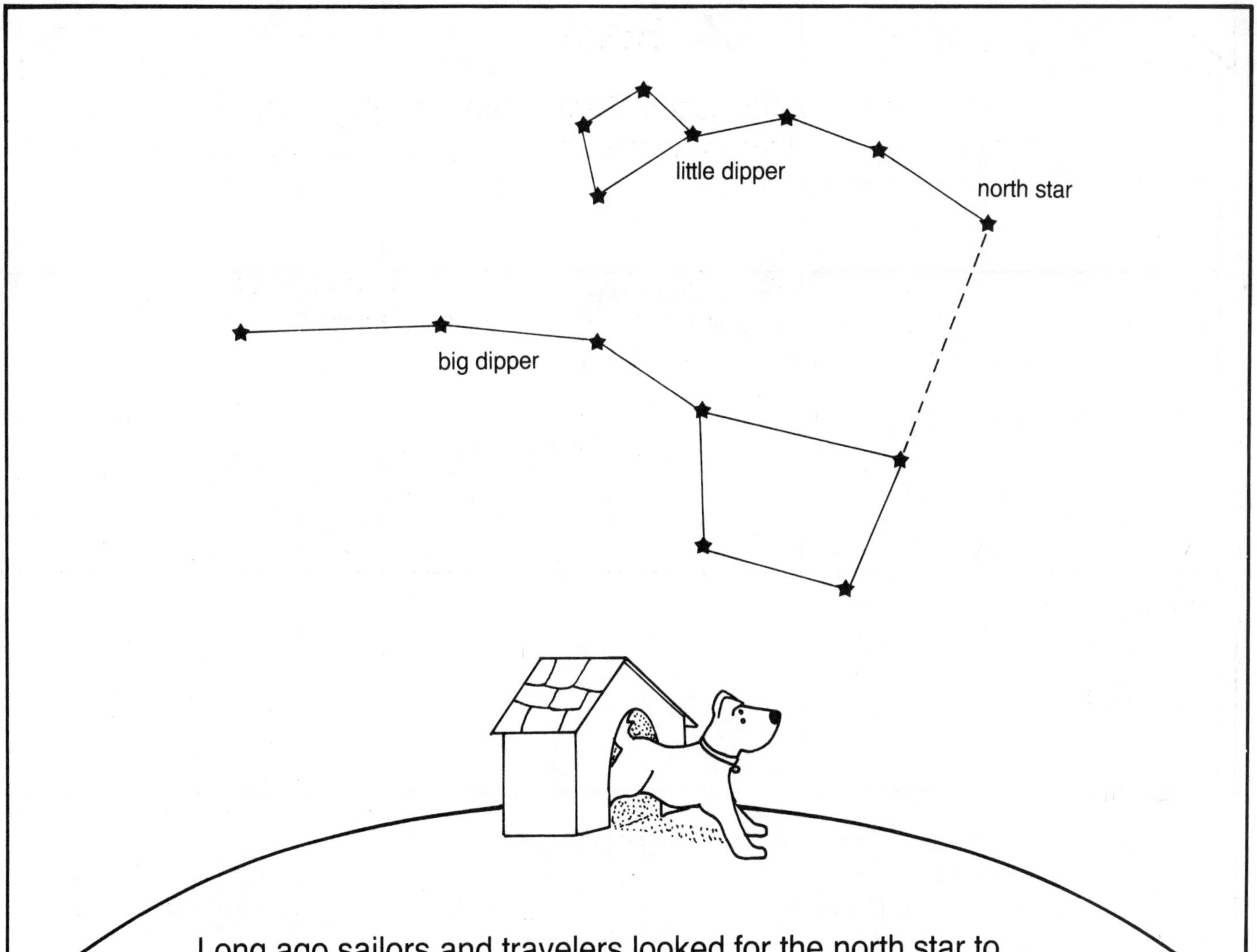

Long ago sailors and travelers looked for the north star to guide them.

Extra: I can find the big dipper in the sky. Yes / No

Name____

Constellations

Here are some of the star pictures in our sky:

Queen on a Throne
(Cassiopeia)

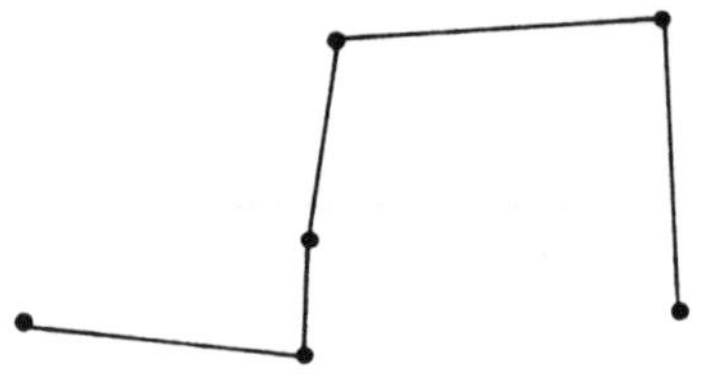

The Dragon
(Draco)

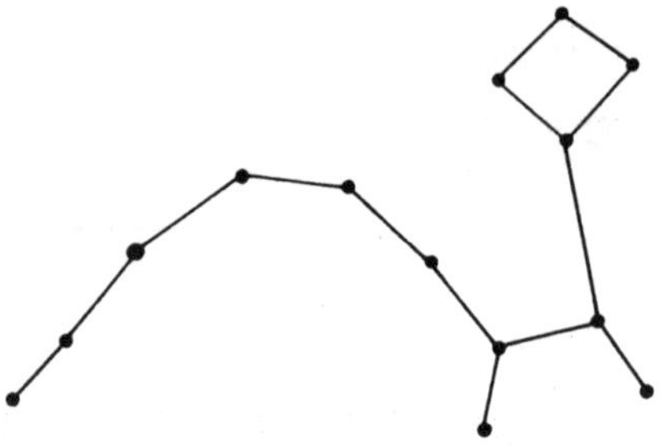

The Hunter
(Orion)

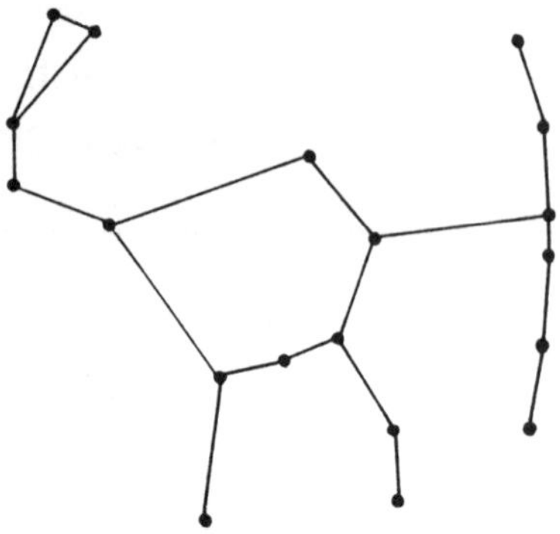

Big Dipper
(Ursa Major)

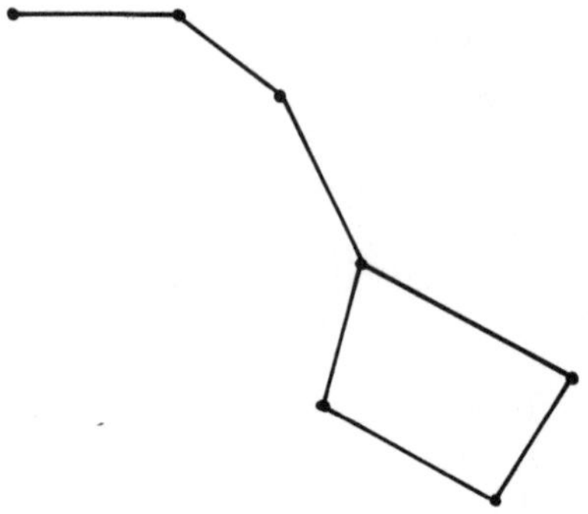

The Winged Horse
(Pegasus)

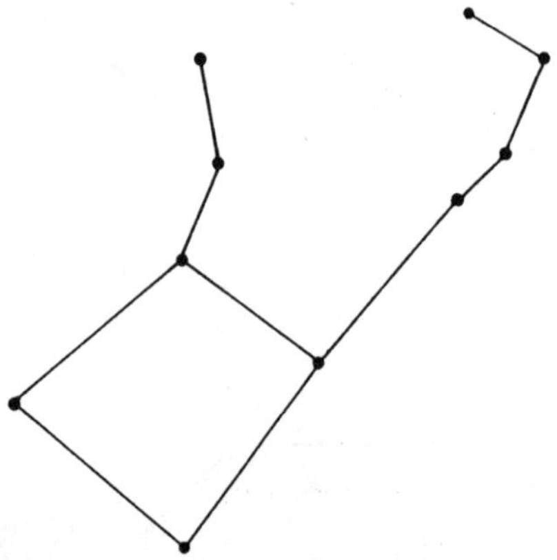

The Swan
(Cygnus)

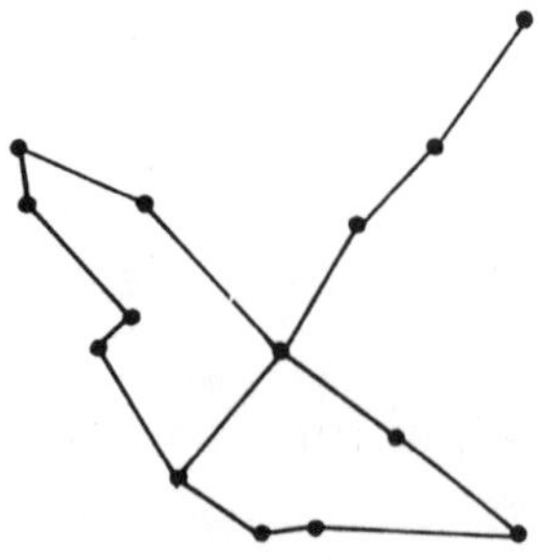

Trace with black crayon:

1.  in Cassiopeia
2. 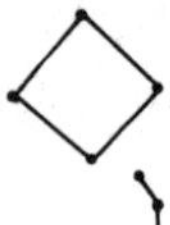 in Draco
3.  in Orion
4. 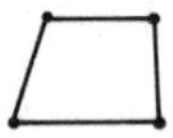 in Pegasus
5. 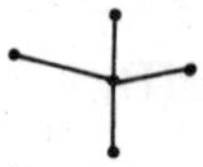 in Cygnus
6.  in Ursa Major

Extra: Draw the little dipper on the back of this paper.

Name ______________________________

Twinkle, Twinkle

Why do stars twinkle? Light comes from the stars in straight lines. When the light gets to the earth, it hits the air that is around the earth. The air makes the light bend and break apart. This makes the stars look like they twinkle.

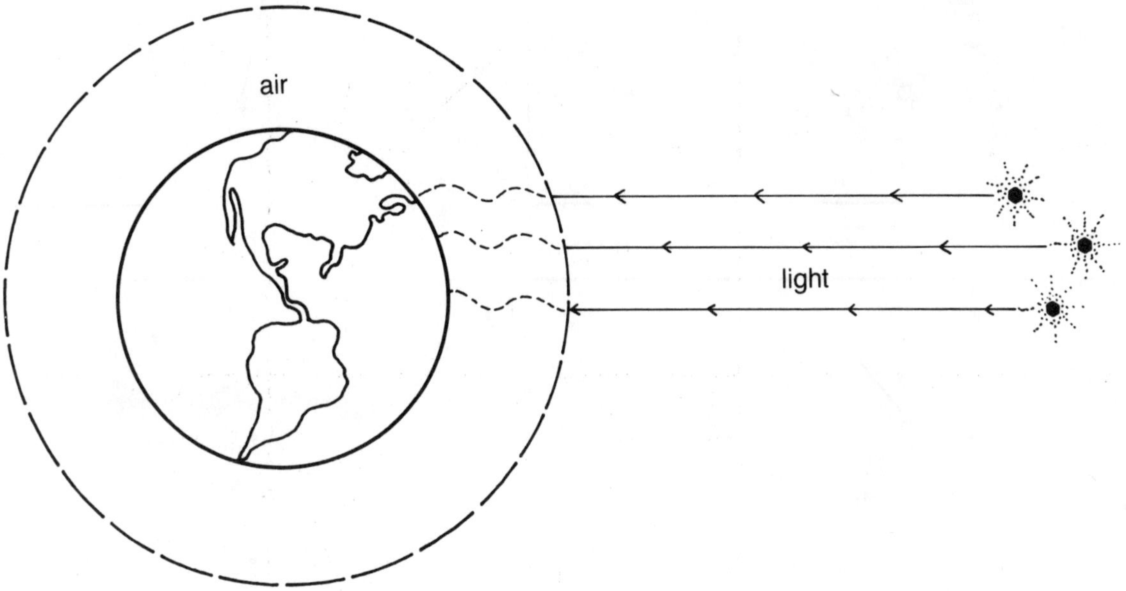

Fill in the blanks:

1. The earth has [ ] around it.

2. [ ] from the stars hits the air.

3. The stars look like they [ ] .

air
light
twinkle

Extra: Say "Twinkle, Twinkle Little Star" to a friend.

Name____

full moon	crescent moon
sun	the dragon
earth	big dipper

What Is It?

Connect the Dots:

A.

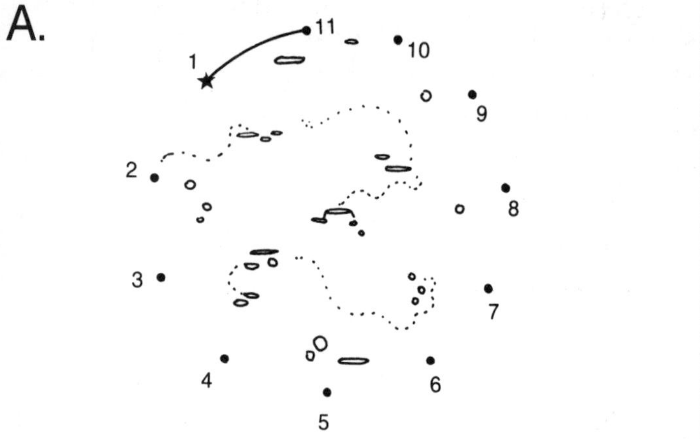

B.

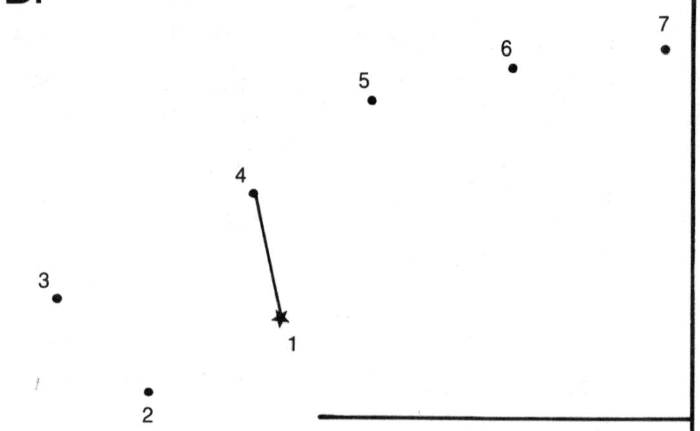

C.

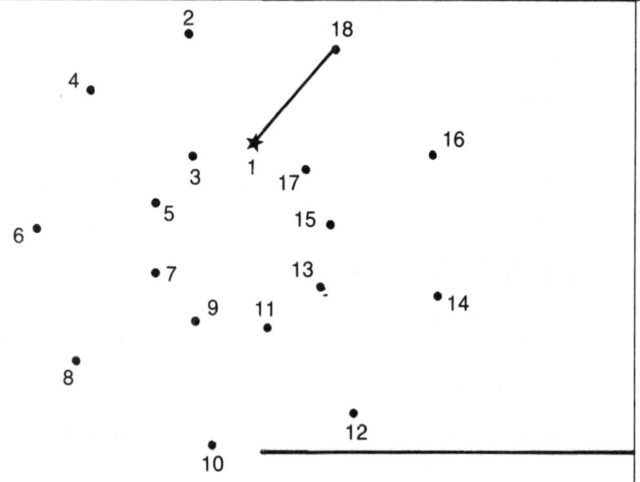

D.

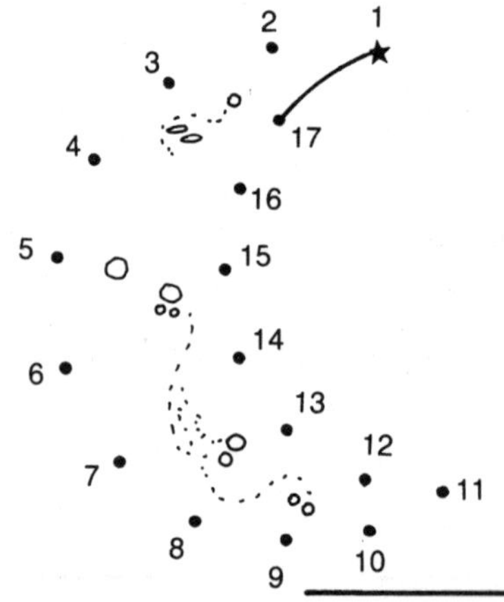

E.

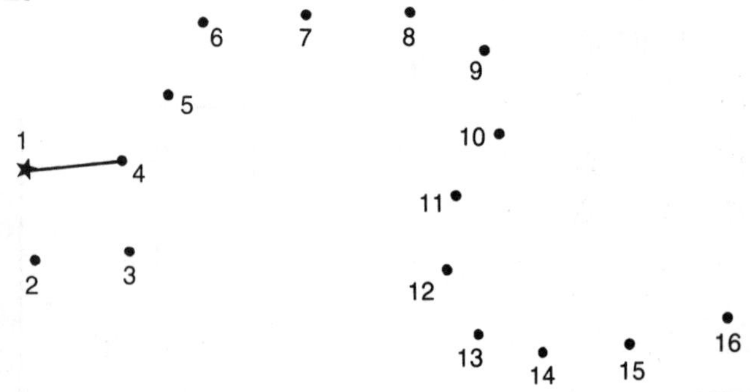

F.

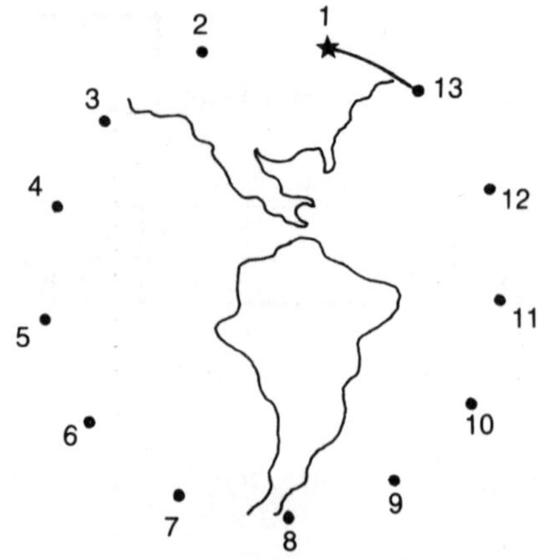

Extra: Color the sun yellow and the earth blue and green.

Name ____________________

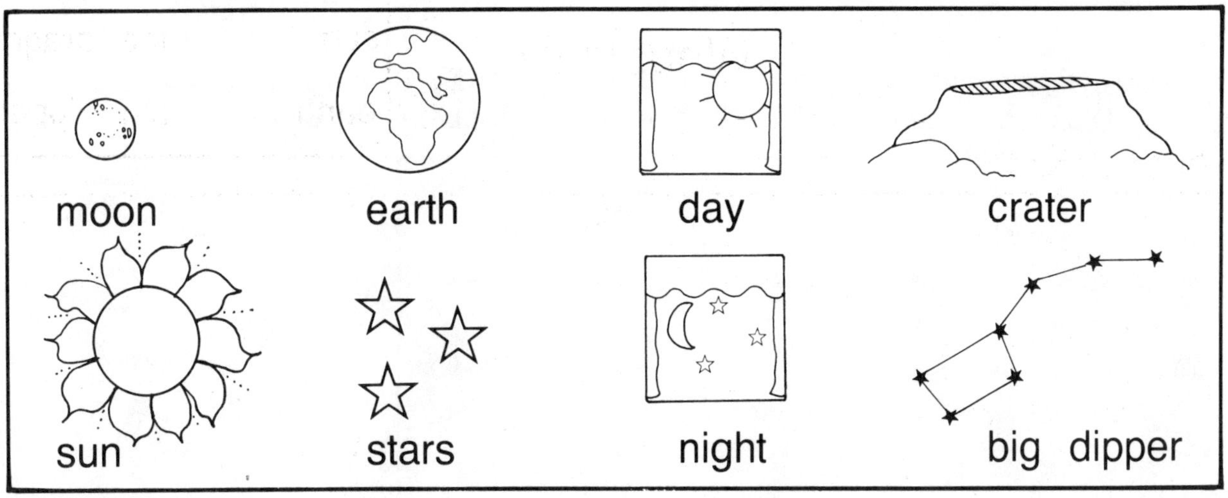

1. I am one of the star pictures you see at night.

2. I am the time when the sky is dark.

3. I am the planet you are on.

4. I am the star you see in the sky all day.

5. I am the time when the sky is full of sunlight.

6. I am a deep hole on the moon.

7. You will see me if you look in the sky at night. I am big and bright.

8. We twinkle in the sky at night.

Can you find the answers here?

A	N	O	O	N	D	R	A	C
B	I	G	D	I	P	P	E	R
U	G	A	A	S	T	E	A	A
S	H	T	Y	S	U	N	R	T
R	T	M	O	O	N	U	T	E
S	T	A	R	S	T	H	H	R

Extra: Make up a riddle about the stars. Tell it to a friend.

Run on tag. Laminate or cover with clear contact paper. Cut playing cards apart. Each player needs a "Big Dipper" card. You will need a box of gummed stars.

Use the poster to review:

Moon	Earth	Orion
Sun	Draco	Pegasus
Cygnus	Pisces	Cassiopeia

The Big Dipper

Rules: Draw a card. If you can name what is on the card, you earn a star for your Big Dipper card. The first player to make the Big Dipper is the winner.

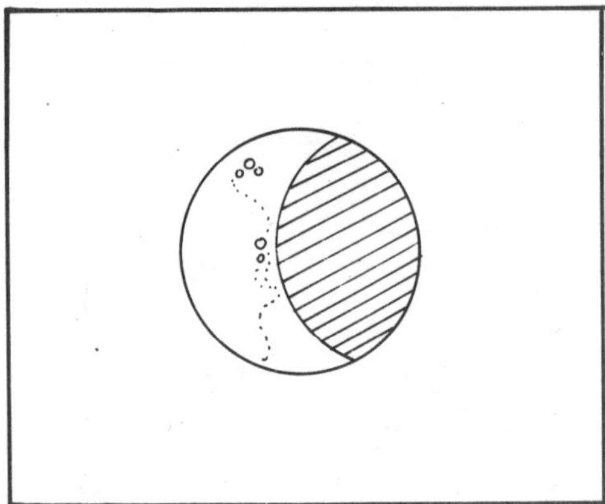
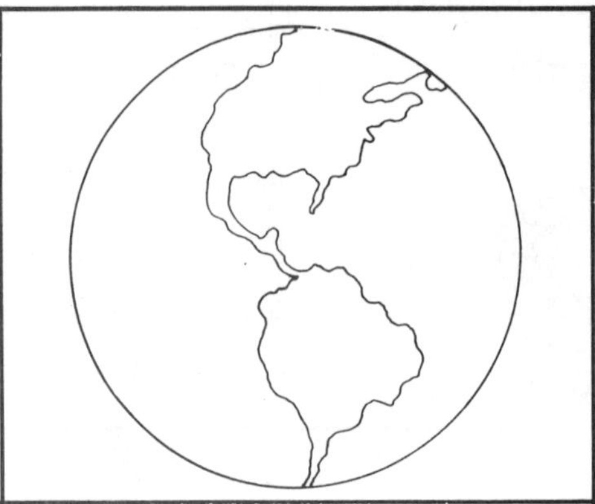
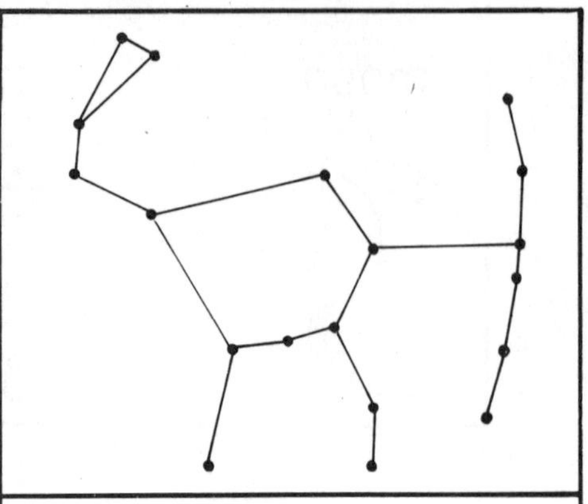
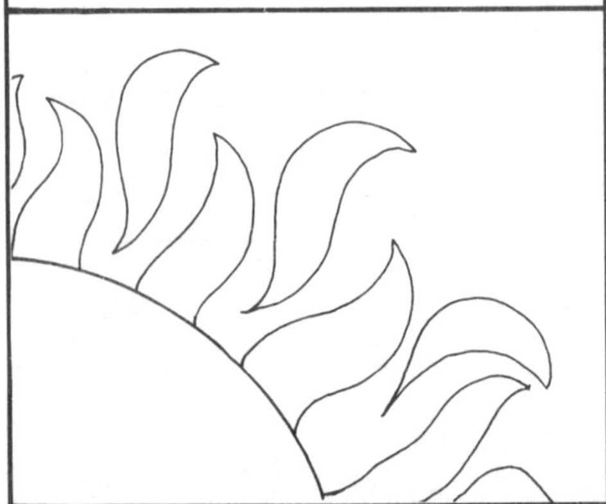
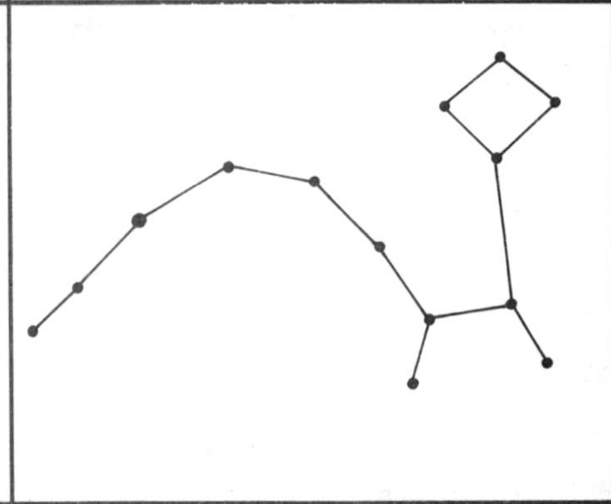
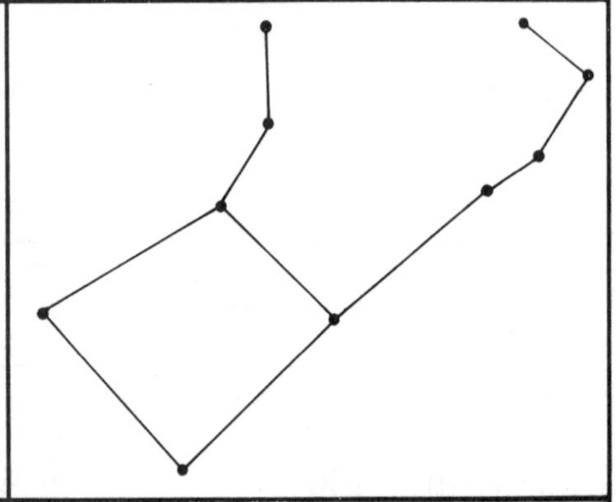
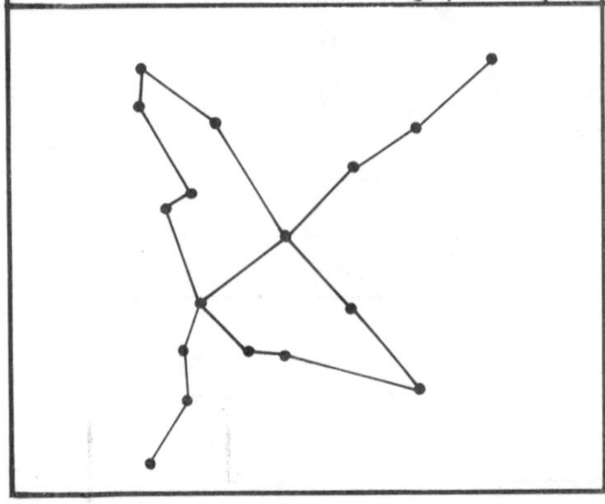
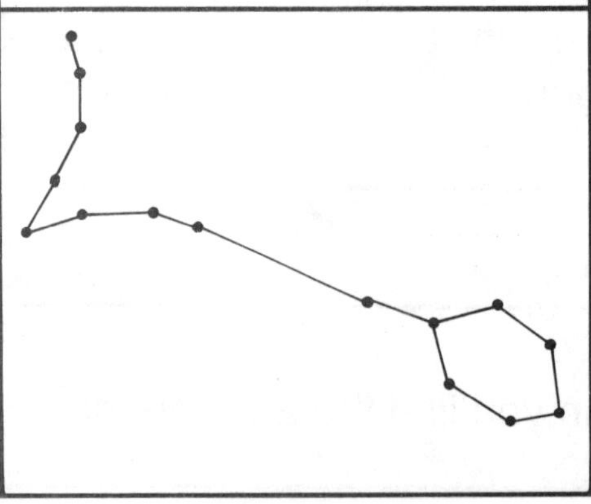
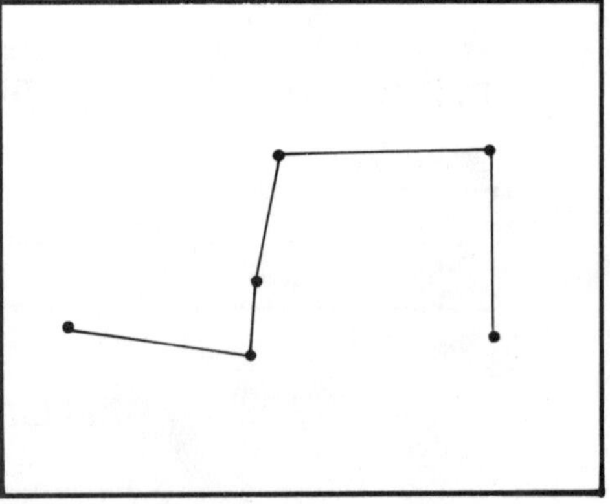

The Big Dipper

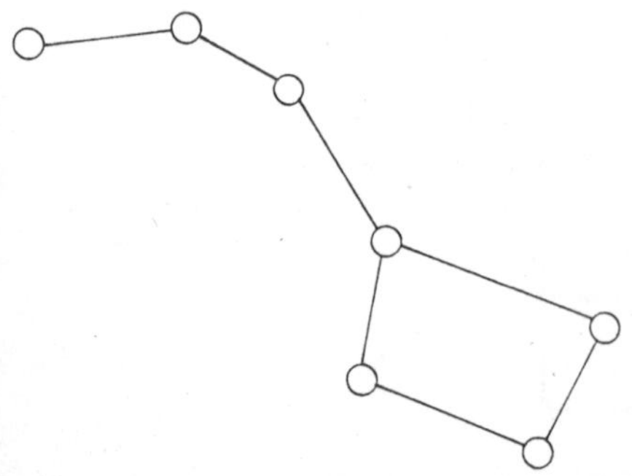

The Big Dipper

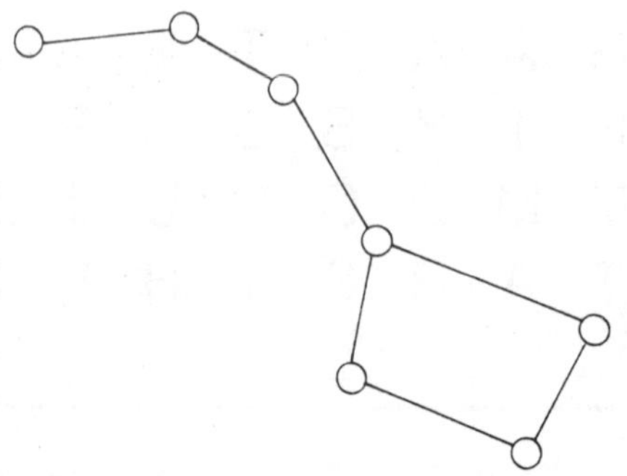

The Big Dipper

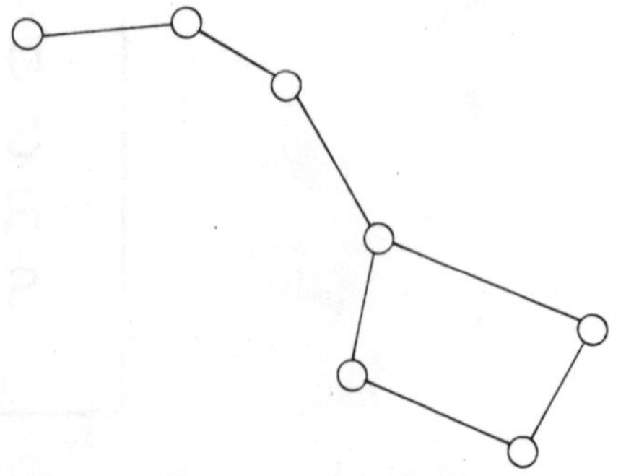